The Adventures of Tom Sawyer

Mark **Twain**

Illustrated by **Paolo D'Altan**

Adaptation and activities by **Sally M. Stockton**

Editor: Victoria Bradshaw
Design and art direction: Nadia Maestri
Computer graphics: Carla Devoto, Simona Corniola
Picture research: Laura Lagomarsino

© 2011 Black Cat

First edition: January 2011

Picture credits:
Cideb Archive; The Mark Twain House & Museum: 4,6; VisionsofAmerica /
Joe Sohm / Getty Images: 26, 28; © Kushnirov Avraham / Dreamstime.com:
58; © Bettmann/CORBIS: 59.

We would be happy to receive your comments and suggestions, and give
you any other information concerning our material.
http://publish.commercialpress.com.hk/blackcat/

CISQ — CISQ CERT
TEXTBOOKS AND
TEACHING MATERIALS
The quality of the publisher's
design, production and sales processes has
been certified to the standard of
UNI EN ISO 9001

ISBN 978 962 07 0450 5   Book + Special CD-ROM

The CD contains an audio section (the recording of the text) and a CD-ROM section
(additional games and activities that practice the four skills).
– To listen to the recording, insert the CD into your CD player and it will play as
  normal. You can also listen to the recording on your computer, by opening your
  usual CD player program.
– If you put the CD directly into the CD-ROM drive, the software will open
  automatically.

SYSTEM REQUIREMENTS for CD-ROM	
PC:	**Macintosh:**
• Pentium III processor	• Power PC G3 or above
• Windows 98, 2000 or XP	(G4 recommended)
• 64 Mb RAM (128Mb RAM recommended)	• Mac OS 10.1.5
• 800x600 screen resolution 16 bit	• 128 Mb RAM free for the application
• 12X CD-ROM drive	
• Audio card with speakers or headphones	
All the trademarks above are copyright.	

Contents

The text is recorded in full.

 These symbols indicate the beginning and end of the passages linked to the listening activities.

Mark Twain (1835-1910)

Mark Twain

 Mark Twain's real name was Samuel Langhorne Clemens. He was born on 30 November 1835 in Florida, Missouri, USA. When he was four years old he and his family went to live in Hannibal, Missouri, a town on the Mississippi River. He grew up on the river and a lot of the adventures in his books happened to him when he was a boy.

When he was twenty-two years old he started working as a steamboat [1] pilot on the Mississippi River. He traveled up and down

1. **steamboat** :

the long river for four years. Then in 1861 he decided to go to Nevada and California: during the 1860s a lot of Americans went to the West to start a new life.

He liked writing and he became a journalist for the *Morning Call* in San Francisco, California. He started to use the name Mark Twain for his writing. In 1865 he wrote the short story "The Celebrated Jumping Frog of Calaveras County" and it was a great success. He became a famous writer and traveled to Hawaii and Europe.

He met Olivia "Livy" Langdon in 1868 and married her in 1870. They had a son and three daughters and lived in Hartford, Connecticut. Here he wrote many of his famous books: *The Adventures of Tom Sawyer* (1876), *The Prince and the Pauper* (1880), *Life on the Mississippi* (1883), *Adventures of Huckleberry Finn* (1884) and *A Connecticut Yankee in King Arthur's Court* (1889). Twain also wrote many funny short stories.

Mark Twain's books became popular everywhere. People liked his way of writing, and many called him the father of American literature. He changed the American way of writing with his simple, funny language.

He died on April 21, 1910 at the age of seventy-four.

1 **COMPREHENSION CHECK**
Choose the correct answer in *italics*.

1 Mark Twain started working as a *journalist/teacher* in San Francisco.
2 He was born in *Florida/Hannibal*, Missouri in 1835.
3 When he was a boy he grew up *on the sea/on the river*.
4 He was steamboat pilot on the *Missouri/Mississippi* River.
5 He changed his name when he got to *Connecticut/California*.
6 In 1865 he wrote a popular *short story/adventure book*.
7 He wrote many of his famous books in *Hartford/San Francisco*.
8 Mark Twain's books became popular in *America/everywhere*.

PROJECT ON THE WEB

Let's visit Mark Twain's house!

Connect to the Internet and go to www.blackcat-cideb.com. Insert the title or part of the title of the book into our search engine. Click on the internet project link. Click on the relevant link for this project.

Work with a partner. Click on *Visitor Information* and answer these questions.

1 What is the address of the Mark Twain House?

2 What are the visitors' hours?

3 On what days is it closed?

4 How much does it cost for adults, older children and children under six?

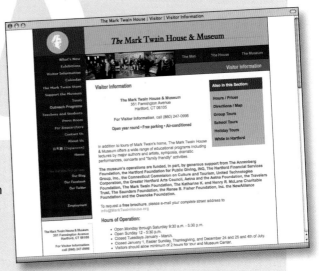

Now click on *The House* and *Map of Floors*. Click on the rooms of the map and take a virtual tour. Which room do you think was the most interesting and why?

Back row: **Aunt Polly, Muff Potter, Injun [1] Joe.**
Front row: **Huckleberry Finn, Becky Thatcher, Tom Sawyer.**

1. **Injun** : Indian, a native American. Injun was the local pronunciation of Indian.

Setting

This story take place along the Mississippi River during the late 1800s. The places in *italics* are fictional. The real name of St Petersburg is Hannibal. This is the town where Mark Twain grew up.

BEFORE YOU READ

1 VOCABULARY
Match the words with each picture.

A fence   B tie   C brush   D kite   E rat

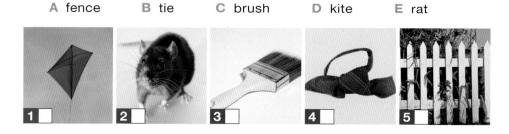

1 ☐   2 ☐   3 ☐   4 ☐   5 ☐

CHAPTER ONE

Painting the Fence

Tom! Tom!"

There was no answer.

"Where is that boy?" said Aunt Polly, looking everywhere. She looked under the bed but she only found the cat.

"Tom! Where are you" she cried loudly. She opened the door and looked in the garden.

Then she heard a noise behind her. A small boy ran past and she stopped him with her hand.

"What are you doing, Tom?" she asked.

"Nothing," answered Tom, looking at his feet.

"Nothing!" cried Aunt Polly. "Look at your hand and your mouth. I told you many times not to eat the jam."

"Oh, Aunt Polly, look behind you!" said Tom.

The old lady looked and Tom ran away. Aunt Polly was surprised and then she laughed.

"I'll never learn," she said to herself. "Tom always plays tricks [1] on me and I never learn. He plays new tricks every day. I love Tom, but it's not easy to look after him. Poor Tom, he's my sister's child and she's dead."

She stopped for a moment and then continued, "Well, tomorrow is Saturday and there's no school so Tom must work. He hates working but he must learn to do it."

Tom lived in the village of St Petersburg, on the Mississippi River in Missouri, with his Aunt Polly and his brother Sid. The summer evenings were long and Tom liked walking around the village. One evening he saw a big boy in front of him. He didn't know him and this surprised Tom because he didn't see new people often. This boy had very nice, expensive clothes. "He's got shoes, a new shirt and a tie. And it's not Sunday," Tom thought. "My clothes are old and ugly and I don't have any shoes."

Tom looked at him and the big boy looked at Tom. Tom didn't like him and said, "I can fight you and I will win!"

"Really?" said the boy. "Why don't you try?"

"Well, I can," said Tom, looking at the boy.

"No, you can't," said the boy angrily.

"Yes, I can," cried Tom loudly.

There was silence.

"You're afraid of me," said the boy.

"No, I'm not afraid of you," said Tom.

"Yes, you are."

1. **tricks** : jokes.

"No, I'm not."

There was silence for a moment.

Then Tom pushed the boy and the boy pushed Tom. Soon they were on the ground fighting. Tom pulled the boy's hair and hit him hard. They both fought hard. Then the big boy started crying.

"Stop!" he cried. "Stop!"

"Now, that will teach you," said Tom and he walked away.

Tom got home late and his clothes were dirty. When Aunt Polly saw him she thought, "What can I do with this boy? Well, tomorrow is Saturday and there's work for him to do."

Saturday morning was beautiful and sunny. It was summer and the world was happy. Tom sat in front of the long, dirty fence and looked at it. It was thirty yards [2] long and nine feet [3] high. He was very unhappy.

"It's Saturday and I have to paint this long fence," he thought. "All my friends are going to laugh at me. It's going to be a very long day"

Tom put the long brush in the white paint and started painting. He stopped and looked at his work unhappily, and then he continued painting.

After a few minutes he had a clever idea. He saw his friend Ben Rogers in the street. Ben had a big red apple in his hand and he came to look at the fence.

"You're working for your Aunt Polly," said Ben, laughing.

Tom said nothing and continued painting.

"I'm going swimming but you can't come with me," said Ben. "You're working."

2. **thirty yards** : about 30 metres.
3. **nine feet** : about 3 metres.

"Do you call this work?" asked Tom.

"Of course it's work. You're painting a fence," said Ben.

"Maybe it's work but maybe it isn't. I like it!" said Tom happily. "I can go swimming every day, but I can't paint a fence every day."

Ben watched Tom as he painted slowly and carefully. He often stopped and moved back from the fence. He looked at his work and smiled. Ben was suddenly interested in the fence and said, "Can I paint a little, Tom?"

Tom thought for a moment and said, "I'm sorry, Ben. Aunt Polly wants me to do it because I'm very good at painting. My little brother Sid wanted to do it, but he's not good at painting."

"Oh, please, Tom,' said Ben. "Please, let me paint! I'm very good at painting too. Here, you can have some of my apple."

"No, Ben, I can't..." said Tom.

"Then take all of my apple," said Ben.

Tom was happy but he didn't smile. He gave Ben the brush and sat down to eat the apple.

Tom's other friends came by. At first they laughed at him, but soon they all wanted to paint the fence. Billy Fisher gave Tom a kite and Johnny Miller brought him a dead rat. His other friends gave him an old knife, a cat with one eye, an old blue bottle, an old key and other interesting things. His friends painted the fence and Tom now had a lot of interesting things. He took these new things and went home.

"Aunt Polly, can I go and play now?" he asked.

When Aunt Polly saw the beautiful white fence she was very pleased. She gave Tom a big apple and said, "Good work! Go and play, but don't be late for dinner!"

UNDERSTANDING THE TEXT

1 COMPREHENSION CHECK

Read these sentences about Chapter One. Choose the correct answer – A, B or C. There is an example at the beginning (0).

0 Aunt Polly told Tom
- A ☐ not to go in the garden.
- B ☐ to wash his hands and mouth.
- C ☑ not to eat the jam.

1 Tom is Aunt Polly's
- A ☐ nephew.
- B ☐ niece.
- C ☐ son.

2 Tom met a big boy in the village
- A ☐ and they became friends.
- B ☐ and they started fighting.
- C ☐ who had old, ugly clothes.

3 Tom didn't like
- A ☐ working.
- B ☐ fighting.
- C ☐ eating jam.

4 On Saturday morning Tom
- A ☐ went swimming with Ben.
- B ☐ played a trick on his friends.
- C ☐ happily painted the fence.

5 Tom's friends painted the long fence
- A ☐ and they gave him some interesting things.
- B ☐ because they could not go swimming.
- C ☐ because Aunt Polly gave them apples and cakes.

SHE OPENED THE DOOR AND LOOKED IN THE GARDEN

We form the Past Simple of regular verbs by adding **-ed** to the verb.

e.g. work = *worked*, talk = *talked*, help = *helped*

Some verbs end in a **consonant** + **-y**. We remove the *'y' and add -ied*.

e.g. *study = studied, carry = carried, hurry = hurried*

But if verbs end in a **vowel** + **-y**, the -y does not change.

e.g. *stay = stayed, play = played, enjoy = enjoyed*

2 THE PAST SIMPLE: REGULAR VERBS

A Complete the sentences with verbs from the box using the Past Simple.

> start   hurry   want   study   carry   push   laugh   play

1 Tom .... a lot of tricks on Aunt Polly.
2 Tom .... math for an hour and then went outside to play.
3 Ben .... painting the fence at ten o'clock.
4 The boys .... because they were happy.
5 The big boy .... Tom to the ground.
6 Aunt Polly .... home because she was late.
7 Tom .... his bag home.
8 All the boys .... to paint the fence.

B Unscramble the verbs and put them in the crossword in the Past Simple form.

1 YCR
2 TYR
3 TANPI
4 OCYP
5 OLKO
6 NPEO

3 VOCABULARY

Read the descriptions. What is the word for each one? The first letter is already there. There is one space for each other letter in the word.

0	you eat it at breakfast or tea on a piece of bread	j a m
1	you use it to paint	b _ _ _
2	you use it to cut a piece of bread	k _ _ _
3	you use it to open the door	k _ _
4	you wear them on your feet	s _ _ _
5	not beautiful	u _ _ _
6	the hot season of the year	s _ _ _ _

T: GRADE 3

4 SPEAKING – FREE TIME

Tom and his friends like to go swimming in their free time. Use these questions to talk to the class about what you do in your free time.

1 What are three things you do in your free time?
2 How often do you do them?
3 Are they easy, difficult or dangerous to do?
4 Why do you like them?
5 Who do you do them with?

BEFORE YOU READ

1 **VOCABULARY**
Match the words with each picture.

A beetle   B grave   C witch   D ghost   E graveyard

1 ☐ 2 ☐ 3 ☐ 4 ☐ 5 ☐

KET

2 **LISTENING**

Listen to the first part of Chapter Two and choose the correct answer – A, B or C.

1 On Sunday Tom had to
   A ☐ wear different clothes from the other days.
   B ☐ paint the long fence.
   C ☐ stay home all day.

2 The people in church
   A ☐ tried not to laugh.
   B ☐ didn't think Tom's trick was funny.
   C ☐ listened to the reverend.

3 On Monday morning Tom
   A ☐ didn't go to school.
   B ☐ went to Huck's house.
   C ☐ talked to Huck.

4 Huck Finn didn't have a home
   A ☐ so he lived in the school.
   B ☐ but he was happy.
   C ☐ and he was sad.

CHAPTER TWO

Tom Meets Becky

The next day was Sunday. Tom wore his special Sunday clothes. He hated them. He and Sid always went to Sunday school [1] on Sunday morning. Tom didn't like Sunday school and never listened to the teacher.

After Sunday school Tom, Sid and Aunt Polly were in church. This Sunday Tom wanted to play a trick in church and brought a big black beetle in his pocket. When the reverend [2] started speaking Tom took the beetle out of his pocket and put it on the floor.

A little dog in the church saw the big beetle and wanted to play with it. But the beetle bit the dog's nose. The little dog barked and all the people in church looked at it. The reverend was very

---

1. **Sunday school** : religious lessons for children in the church on Sunday.
2. **reverend** : a religious leader in some religions.

surprised. The dog jumped up and down and ran after the black beetle. He made a lot of noise.

Everyone thought it was funny but they tried not to laugh. The reverend continued talking but no one was listening to him and he was angry. Everyone was interested in the dog and the black beetle. Tom was happy.

On Monday morning Tom did not want to get up and go to school. Aunt Polly went to his room and shouted, "Tom, get up immediately and get ready for school!"

That morning Tom met his good friend Huckleberry Finn and they walked together. Huck's father was lazy [3] and didn't work and his mother was dead. He had no home; he lived in the streets. He didn't go to school or to church. His clothes were old and dirty. He was free to do what he wanted so he always went fishing and swimming. He was a happy boy.

The mothers of St Petersburg didn't like him because he often used bad language and didn't listen to anyone. But the children of the town liked him a lot. They wanted to be free like him.

"Hello, Huck!" said Tom. "What do you have in the bag?"

"It's a dead cat," said Huck.

"What are you going to do with a dead cat?" asked Tom.

"I want to take it to the graveyard after midnight," said Huck. "A dead cat can call ghosts out of their graves."

"Oh, really?" asked Tom, who was surprised. "Is it true?"

"Well, old Mrs Hopkins told me," said Huck. "She knows about these things because she's a witch."

"Can I come with you?" asked Tom.

"Of course," said Huck, "but aren't you afraid of ghosts?"

3. **lazy** : a lazy person doesn't like working.

"Afraid of ghosts! Of course not!" said Tom. "Come and call me at my window tonight at eleven o'clock."

Tom went to a small, one-room school in St Petersburg. That morning he got there late and the teacher was angry.

"Thomas Sawyer!" said the teacher. "Why are you late *again*?"

Tom looked around the classroom and saw a new girl. She had long blonde hair and blue eyes. She was very pretty and Tom liked her. There was a free chair next to her and Tom wanted to sit there. But how?

Tom thought quickly and said, "I met Huckleberry Finn on the street and stopped to talk to him."

The teacher got very angry and said, "You know you must never talk to that terrible boy! Now go and sit down with the girls!"

All the children laughed at Tom. He sat down next to the new girl and looked at her. He was happy and drew a picture of a house.

"Let me see your picture," she whispered. [4]

Tom showed her the picture.

"It's nice," she said. "Now draw a man."

Tom drew a man near the house. It was a bad picture but the girl liked it.

"You draw well," she said smiling. "I can't draw."

Tom felt happy and said, "I can teach you after school."

"Oh, thank you, Tom!" she said.

"What's your name?" Tom asked.

"Becky Thatcher," she said. "I know your name; it's Tom Sawyer."

---

4. **whispered** : said in a very quiet voice so that no one could hear.

That night Tom and Sid were in bed at half past nine. Sid was soon sleeping but Tom wasn't. He was waiting for Huck. At eleven o'clock he heard Huck outside and got out of bed. He got dressed and quickly went out of the bedroom window.

"Let's go!" whispered Huck, holding the bag with the dead cat. Tom and Huck walked to the graveyard on the hill. It was a dark, scary[5] place with a lot of trees and a lot of graves. The wind made strange noises and dark clouds covered the moon.

"Are the ghosts making the noises?" Tom thought. He was afraid but he didn't say anything.

"Let's find the grave of Hoss Williams," said Huck, looking around.

They walked around the graveyard and soon found Hoss Williams's grave.

"Well, here it is," said Huck. "He died last week."

"Do you think he can hear us?" asked Tom, who was scared.

"I think his ghost can hear us," said Huck.

"Then let's call him *Mr* Williams," said Tom.

"Alright," said Huck. "But everyone called him Hoss."

"Sh!" whispered Tom, his heart was beating[6] fast and he felt cold.

"What is it?" asked Huck.

"Can you hear that noise?" asked Tom. "Look over there, Huck. Oh, no!"

5. **scary** : it frightened them.
6. **beating** : (here) moving.

UNDERSTANDING THE TEXT

1 COMPREHENSION CHECK

Read the paragraphs below and choose the best word (A, B or C) for each space. There is an example at the beginning (0).

On Sunday Tom put on (0) ......his...... Sunday clothes and went to Sunday school. Then he went to church (1) .... Aunt Polly and Sid. He took a black beetle to church (2) .... he wanted to play a trick. He put the beetle on the floor and a dog ran (3) .... it. There was a lot of noise in church. (4) .... thought it was very funny and the reverend was angry.

The next day on the way to school
Tom met his friend Huckleberry Finn,
(5) .... didn't have a home. Huck was
(6) .... a bag with a dead cat inside.
He asked Tom to go to the graveyard at midnight.

Tom got to school late and the teacher was angry. Tom had to sit
(7) .... to a pretty girl called Becky and he liked her a lot.

That night Tom and Huck went to the graveyard and looked
(8) .... Hoss Williams's grave. Tom heard a noise and was afraid.

0	A its	B him	C his
1	A with	B by	C at
2	A because	B why	C that
3	A before	B after	C back
4	A Everyone	B Every	C Anyone
5	A he	B that	C who
6	A carried	B carrying	C carried
7	A next	B by	C with
8	A by	B to	C for

TOM TOOK THE BLACK BEETLE OUT OF HIS POCKET AND PUT IT ON THE FLOOR.

Some verbs have irregular Past Simple forms, e.g. took (take) and put (put).

Here are some other examples:

be = was/were	come = came	eat = ate
go = went	fight = fought	leave = left

2 THE PAST SIMPLE: IRREGULAR VERBS

A Complete the table below with the Past Simple form of the verbs below. All the verbs are in the first two chapters of the story.

1	run	7	say ....
2	tell	8	give ....
3	see	9	bring ....
4	know	10	do ....
5	think	11	make ....
6	have	12	feel ....

B Now use the irregular verbs in the box below to complete the sentences. Put them into the Past Simple.

be   draw   hear   meet   sit   wear

1 Tom .... Aunt Polly calling him
2 Tom .... his best clothes to go to Sunday school.
3 Tom, Sid and Aunt Polly .... in church on Sunday morning.
4 Tom .... his friend, Huck, on his way to school.
5 Tom .... next to Becky Thatcher.
6 He .... some pictures for Becky.

3 ODD ONE OUT

A **Circle the word that does not belong to each group. Say why.**

1	tie	shoes	shirt	jacket
2	jam	apple	orange	banana
3	lunch	dinner	meat	breakfast
4	swimming	reading	running	dancing
5	village	city	school	town
6	rat	cat	dog	beetle

B **Now use the odd words to complete the sentences.**

1 Becky liked .... books in her room.
2 Tom didn't have any .... on his feet.
3 The dog wanted to play with the ....
4 Aunt Polly told Tom not to eat the ....
5 Tom didn't like going to Sunday ....
6 Huck had a big piece of .... for lunch.

KET

4 WRITING

Complete Becky's diary. Write ONE word for each space. There is an example at the beginning (0).

Today was (0) ......my...... first day at the new school
(1) .... St Petersburg. It is a one-room school
(2) .... it is near my house.
I met a new boy at school. His name (3) .... Tom
Sawyer and I like (4) ....
He was late to school and the teacher got angry. Tom sat next
to (5) .... He is very friendly. Tomorrow after school he
is coming (6) .... my house (7) .... four o'clock.
He is going to teach me how to draw. He can have ice-cream
and cake (8) .... me in the garden.

The One-Room School

5 Tom Sawyer, Becky Thatcher and the other children of St Petersburg went to a one-room school. There were many one-room schools in villages and small towns in the United States during the eighteenth and nineteenth centuries. During the late 1800s there were about 190,000 one-room schools, and today there are about 400.

What was the one-room school like?

The school looked like a small house with one big room. On top of the roof of the school there was a school bell. [1] Inside the big room there were small wooden desks and chairs for the students. The teacher had a desk at the front of the room. Behind the teacher's

1. bell :

desk there was a big blackboard. [2] The students' parents usually made the desks and chairs. Every one-room school had a stove. [3] During the winter the teacher burned wood in the stove to make the room warm. The children who sat near the stove were often too hot, and those who sat near the windows were often too cold.

The girls sat on one side of the room and the boys sat on the other. The younger students sat in the front near the teacher and the older ones sat in the back. The youngest students were about six years old, and the oldest were about fourteen or fifteen. There were between six to forty students in the one-room school. The teacher had to teach them all!

There weren't many books or pictures on the walls. Usually there was only a map of the United States.

Teachers did not have an easy job. In the winter they had to get to school very early to put wood in the stove. During the cold winter days they often made hot soup on the stove for the students' lunch. Sometimes they had to clean the room too!

What was the school day like?

Some children walked to school and others rode a horse. The school day began at 8 a.m. and ended at 4 p.m. The first subject was always reading. Then there were games for fifteen minutes. The second subject was math and then writing. After an hour's break for lunch it was time for spelling and grammar, and then history. The last subject was geography. The teacher taught the younger students first and then the older students. The older students often helped the younger ones with reading and math.

There was no homework because most students had to work on the family farm when they got home.

2. **blackboard** :   3. **stove** :

One-room schools today

An interesting example of a one-room school today is the Monhegan Island School. It is on Monhegan Island off the coast of the state of Maine. Fifty people live in the village and there are only seven students in the school. Their ages are from five to twelve.

Another example is the Death Valley Elementary school in California. Death Valley is America's hottest desert and few people live there. Eleven students go to the Death Valley Elementary School and their ages are from five to nine.

Teachers and students today like the one-room school because it's like a big family and the students get a lot of attention. [4]

1 **COMPREHENSION CHECK**

Are the following sentences true (T) or false (F)? Correct the false ones.

		T	F
1	Today there are no one-room schools in the United States.	☐	☐
2	During the winter there was no heating in the one-room school.	☐	☐
3	The boys and girls did not sit together.	☐	☐
4	The older students did not sit near the teacher.	☐	☐
5	Teachers didn't ever cook lunch for the students.	☐	☐
6	Lessons were over at 4 p.m.	☐	☐
7	After lunch the boys and girls studied reading.	☐	☐

4. **attention** : when you show great interest to someone.

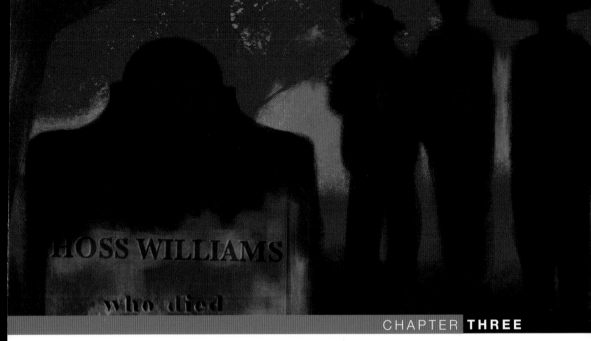

HOSS WILLIAMS

who died

CHAPTER THREE

Ghosts!

Ghosts!" said Huck. "I can see ghosts! They're coming here and I'm really scared!"

"Can ghosts see us?" asked Tom.

"Sure, ghosts can see everything," answered Huck. "Oh, why did I come here?"

"Don't be afraid," said Tom. "Let's not make any noise."

The three ghosts moved quietly in the graveyard. They came close to Tom and Huck.

"Tom!" whispered Huck. "They're not ghosts, they're people. And one of them is Muff Potter. I know his voice."

"You're right, Huck," said Tom. "And there's Injun Joe and Dr Robinson. But what are they doing here?"

"I know," said Huck. "They're grave robbers. [1] They want to steal a dead body from a grave."

"Ugh! But why?" asked Tom, who was very surprised.

1. **robbers** : thieves, people who steal things from other people.

"Dr Robinson wants dead bodies because he cuts them and studies them," said Huck. "My father told me about him."

The three men were at Hoss Williams's grave. Injun Joe and Muff Potter started digging. [2] Soon the grave was open. They found the dead body and slowly pulled it out of the ground.

"Well, Doctor," said Muff, "if you want us to take the body to your house you must give us five dollars."

"What!" said Dr Robinson angrily. "I paid you this morning. I'm not giving you any more money!"

"I want more money, Doctor," said Injun Joe. "Five years ago I came to your father's house. I was hungry and I asked you for something to eat. Your father gave me nothing. I still remember that. Now you must give me more money."

Injun Joe was angry and took the doctor's arm. The doctor hit Injun Joe and they both fell to the ground.

"Don't hit my friend!" cried Muff Potter. Muff and Dr Robinson started fighting.

Everything happened very quickly. Dr Robinson hit Muff Potter on the head. Muff fell to the ground unconscious [3] and Injun Joe took Muff's knife. He saw Muff on the ground and he killed Dr Robinson with the knife. The doctor fell on top of Muff and covered him with blood. [4]

Injun Joe looked at the two men on the ground. First he stole money from the dead doctor's pockets. Then he put the bloody knife into Muff's right hand.

After a few minutes Muff moved a little and opened his eyes. He pushed the doctor's body away and looked at the knife in his hand.

"What – what happened, Joe?" he asked slowly.

2. **digging** :

3. **unconscious** : not awake.

4. **blood** :

"Something very bad, Muff. You killed the doctor! Why did you kill him?" said Injun Joe.

"I didn't kill him!" said Muff, who was very confused. [5] "The doctor hit me on the head and I fell... Now I can't remember anything after that. Tell me, Joe, what happened?"

"You fought with the doctor," said Injun Joe. "He hit you on the head and you fell to the ground. Then you got up, took your knife and killed him."

"I don't understand," cried Muff. "I never fight with a knife. I didn't want to kill Dr Robinson. He was young and a very good doctor: he had a future. Oh, this is terrible! Joe, don't tell anyone, please!"

"Don't worry. I won't tell anyone," said Injun Joe. "But now you have to leave this graveyard. Quickly. Go!"

"Thank you, Joe," said Muff. "You're a true friend."

Muff Potter ran away and Injun Joe watched him. Then he carefully put Muff's knife near the doctor's body and left the graveyard.

Tom and Huck saw everything that happened and they were terrified. They quietly moved away and ran out of the graveyard and back to the village. They got to an old house and decided to hide there.

"What are we going to do?" asked Tom. "We saw everything. Injun Joe killed the doctor."

"What can we do?" said Huck. "We can't tell anyone. I'm afraid of Injun Joe. He's dangerous. Do you want a knife in your heart?"

"I'm afraid of him, too," said Tom. "You're right, Huck, we can't tell anyone about Injun Joe and what happened tonight. But poor Dr Robinson, and poor Muff!"

"Promise not to tell anyone!" said Huck, looking at Tom with his big eyes.

"I promise," said Tom. "I promise!"

5. **was... confused** : didn't understand what was happening.

UNDERSTANDING THE TEXT

1 COMPREHENSION CHECK

Are these sentences "Right" (A) or "Wrong" (B)? If there is not enough information to answer "Right" or "Wrong", choose "Doesn't say" (C). There is an example at the beginning (0).

0   Huck and Tom were afraid because they thought they could see ghosts.
   (A) Right   B Wrong   C Doesn't say

1   The three ghosts were men: Muff Potter, Injun Joe and Dr Robinson.
   A Right   B Wrong   C Doesn't say

2   The grave robbers wanted to steal money from the graves.
   A Right   B Wrong   C Doesn't say

3   Dr Robinson was a very rich man.
   A Right   B Wrong   C Doesn't say

4   Injun Joe asked Dr Robinson for more money.
   A Right   B Wrong   C Doesn't say

5   Muff Potter used his knife to kill Dr Robinson.
   A Right   B Wrong   C Doesn't say

6   Injun Joe took money from the dead doctor's pockets.
   A Right   B Wrong   C Doesn't say

7   Muff always fought with a knife.
   A Right   B Wrong   C Doesn't say

2 PREPOSITIONS

Complete the sentences with a preposition from the box. One preposition can be used more than once.

on	for	near	at	in	from	with

1   Tom and Huck were .... Hoss Williams's grave.
2   The men started fighting .... a big knife.
3   Aunt Polly received an important letter .... her brother.
4   Tom painted the fence .... a long brush and some paint.
5   "This blue bottle is .... you, Tom," said his friend Jason.
6   Tom and his friends had lunch .... noon.
7   School was over .... July.
8   "Put the bag .... the table," said Aunt Polly.

3 CHARACTERS

Read the descriptions of the characters and complete the crossword puzzle.

Across

2 He doesn't have a family.

5 He wanted dead bodies to study.

7 She is new at school.

8 She looks after her nephews.

Down

1 He thinks he killed someone.

3 He lives with his older brother and his aunt.

4 He likes playing tricks on people.

6 He killed someone.

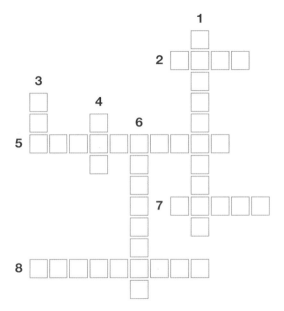

4 DISCUSSION – GHOSTS

Tom and Huck went to the graveyard on the hill with a dead cat in a bag. They wanted to call ghosts out of their graves, but they were afraid of ghosts. What do you think? Discuss these questions.

1 Do ghosts really exist?

2 What do they look like?

3 Do they make noise?

4 Are they good or bad?

5 Do you know any story about ghosts? Tell the class about it.

BEFORE YOU READ

1 VOCABULARY

Match the words below with each picture.

A frying pan   B jail   C raft   D pirate   E treasure

1 ☐   2 ☐   3 ☐   4 ☐   5 ☐

KET

2 LISTENING

 Listen to the first part of Chapter Four and choose the correct answer – A, B or C.

1 The people of St Petersburg talked about
   A ☐ Injun Joe.
   B ☐ Muff Potter.
   C ☐ Dr Robinson.

2 Tom and Huck knew that
   A ☐ Muff did not kill Dr Robinson.
   B ☐ Dr Robinson was rich.
   C ☐ Injun Joe was a good friend of Dr Robinson.

3 Tom and Huck kept the secret
   A ☐ because they were afraid of Injun Joe.
   B ☐ because they were afraid of Aunt Polly.
   C ☐ and they forgot what they saw.

4 Tom wanted to go to Jackson's Island
   A ☐ with Huck.
   B ☐ with Becky.
   C ☐ with Joe and Huck.

CHAPTER FOUR

Jackson's Island

The next day everyone was talking about poor Dr Robinson. The sheriff [1] found Muff Potter's knife near the body of the doctor. He put Muff in St Petersburg's small jail.

Tom and Huck were worried. "We saw Injun Joe kill the doctor," said Tom. "Muff didn't kill him. Poor Muff!"

"I know," said Huck. "I'm very sorry for Muff Potter, too. But we mustn't say anything. Remember, Injun Joe is dangerous — very dangerous."

"We must keep this a secret," said Tom sadly.

At night Tom had bad dreams about Dr Robinson, Injun Joe and Muff Potter. He kept the terrible secret but he couldn't forget what he saw. Aunt Polly was worried about him. She gave him a lot of

---

1. **sheriff** : a policeman in some parts of the United States.

different medicines, but Tom did not feel better. He was unhappy at school too, and Becky Thatcher didn't talk to him anymore.

"What a horrible life!" thought Tom. "No one loves me, and I'm very unhappy."

Then it was summer and school was over. Tom and his friend Joe Harper went fishing on the Mississippi River. They talked and watched the big steamboats go up and down the river. One day Tom said, "Let's do something different and exciting!"

"That's a great idea," said Joe, who was always ready to have some fun. "But what shall we do?"

"Let's go and live on Jackson's Island," said Tom excitedly. "We can be pirates. A pirate's life is exciting."

Jackson's Island was a small island near St Petersburg. It was on the Mississippi River and no one lived there.

"Huck Finn can come with us," said Tom. "Remember, Joe, don't tell your mother, father or anyone about our adventure. Go home and bring some food. Let's meet here at midnight." Tom and Joe were excited.

At midnight the three boys met on the river. Tom brought some meat, Joe brought some bread and Huck brought a frying pan. They were ready to start their exciting adventure.

They found an old raft and they went down the river to Jackson's Island. When they got to the island they made a fire under a big tree and cooked some meat.

"This is great fun!" said Joe.

"We're free and we can do anything we want!" said Tom.

What do pirates do?" asked Huck.

"They go on ships and steal the treasure," said Tom. "Then they go to an island and hide it in a secret place."

The three boys talked about great adventures and then slept

under the stars during the warm summer night. The were very happy.

The next morning was sunny and hot, and the boys went swimming in the river. Then they went fishing and cooked the fish for breakfast. They were hungry and ate it quickly.

After breakfast they explored the island and went swimming again. In the afternoon they sat around the fire and ate some meat and bread. Suddenly Tom said, "Can you hear a strange noise? Listen."

"I can hear it," said Joe. "What is it?"

"Let's go and see," said Huck.

They ran to the river and saw two big steamboats and a lot of small boats near them.

"What's happening?" said Joe. "Every boat from St Petersburg is out on the river and there are two big steamboats too."

"They're looking for a dead body," said Huck. "The same thing happened last summer when little Bill Turner fell into the river and drowned." [2]

"Who are they looking for this time?" asked Joe, who was worried.

Tom thought for a moment and said, "I know! They're looking for *us*! They think we drowned."

The three boys laughed and felt very important.

"Everyone in St Petersburg is looking for us and talking about us," said Tom happily. "We're really famous!"

Tom, Huck and Joe were having a wonderful time. They felt like real pirates on Jackson's Island. In the evening the boats and steamboats went away. The boys went fishing again and cooked

2. **drowned** : died in the water.

some fish for dinner. Huck and Joe slept under the stars, but Tom was worried.

The next morning when Huck and Joe woke up, Tom wasn't there.

"Tom isn't here," said Joe. "Where is he?"

"I don't know," said Huck, looking around.

After a few minutes Huck said, "Look, Joe! Tom's swimming in the river. He's coming back to the island."

Tom got out of the water and told them his story.

"I couldn't sleep last night," he said. "I was thinking about Aunt Polly. So I went home but no one saw me. I saw Aunt Polly and your mother, Joe. They were both crying and very sad. Everyone thinks we're dead. I heard some other interesting things, too."

"What did you hear?" asked Huck. "Tell us."

"Well, there will be a funeral[3] for us on Sunday at the church," said Tom. Huck and Joe looked at him with big eyes. "And now I have a great idea. Listen…"

Tom told his friends about his great idea. They liked it and laughed.

Sunday was the day of the funeral. There were no happy faces in St Petersburg. Everyone in the village was in the small church. Aunt Polly, Sid and Joe Harper's family were all dressed in black. The reverend said many kind words about the three boys. The boys' families cried a lot. Becky Thatcher cried too. Everyone was sad.

Suddenly there was a noise at the church door. The reverend looked up and stopped speaking. Everyone turned around and

---

3. **funeral** : a ceremony when someone dies to say goodbye.

looked with their mouths open. The three boys slowly walked into the church. Tom was first, then Joe and then Huck.

Everyone was silent for a moment. They couldn't believe their eyes. How was this possible? They were in the church for the boys' funeral, but instead…

Suddenly Aunt Polly, Sid and Joe's mother got up and ran to the boys. They kissed Tom and Joe. Aunt Polly cried and then laughed.

Poor Huck didn't know what to do. No one looked at him and no one kissed him. He was sad and started moving away but Tom stopped him.

"Aunt Polly, it's not right," said Tom. "Huck's my good friend. Isn't anyone happy to see him?"

"Oh, Tom, you're right," said Aunt Polly, looking at Huck. "We're all happy to see him."

She smiled at Huck and kissed him. He was happy. Tom was very proud [4] of his great idea.

Then the reverend looked at the three boys and said, "This is a wonderful surprise. Tom, Huck and Joe are here with us and they're well."

Everyone in the small church laughed.

"Let's be happy and sing!" said the reverend happily.

4. **proud** : happy and satisfied.

UNDERSTANDING THE TEXT

1 COMPREHENSION CHECK

Complete the following sentences (1-8) with their endings (A-I). There is an example at the beginning (0).

0 | G | The sheriff put Muff Potter in jail
1 | ☐ | Tom and Huck were afraid of Injun Joe
2 | ☐ | No one lived
3 | ☐ | Tom, Huck and Joe went to Jackson's Island
4 | ☐ | A lot of boats from St Petersburg were on the river
5 | ☐ | Tom went back to St Petersburg at night
6 | ☐ | Everyone thought the boys were dead
7 | ☐ | On the day of the funeral
8 | ☐ | When Aunt Polly, Sid and Joe's mother saw the three boys

A | and they were looking for the three boys.
B | they were happy and kissed them.
C | on a raft.
D | and they planned a funeral for them.
E | because he was dangerous.
F | and saw Aunt Polly and Joe's mother crying.
G | because his knife was near Dr Robinson's body.
H | Tom, Huck and Joe walked into the church.
I | on Jackson's Island.

KET

2 WRITING

You are Tom. Complete your diary. Write ONE word for each space. There is an example at the beginning (0).

Sunday, July 10

Today was (0) ......an...... exciting day. Huck, Joe (1) .... I left Jackson's Island early (2) .... the morning. We took the old raft and went back (3) .... St Petersburg. Then (4) .... went to the village (5) .... to see our funeral. Everyone in the church (6) .... sad. The reverend was (7) .... some kind words about us. Suddenly we (8) .... the door and walked into the church. Everyone turned (9) .... and looked at us. Aunt Polly, Joe's mother and Sid ran to us and (10) .... us. We were all very happy.

3 VOCABULARY – THE WEATHER

Look at these kinds of weather and write the correct name under each picture.

> foggy   windy   sunny   snowy   cloudy   rainy

1	2	3
4	5	6

4 SPEAKING: THE WEATHER

Talk about the weather. Use these questions to help you.

1   What is your favorite kind of weather?
2   What do you do when the weather is like this?
3   Which kind do you like least?
4   What do you usually do when the weather is like this?
5   What is the weather usually like where you live? In January?
    In April? In July? In October?

PROJECT ON THE WEB

Let's find out more about the Mississippi River!

Connect to the Internet and go to www.blackcat-cideb.com. Insert the title or part of the title of the book into our search engine. Click on the internet project link. Click on the relevant link for this project.

Many of Mark Twain's stories take place on the Mississippi River. It is the longest and most important river in the United States. The Missouri River is part of the Mississippi. Today millions of Americans live, work and travel on this great river. Work with a partner and answer the questions below.

1 How long is the Mississippi River, including the Missouri River?

2 Where does the Mississippi start and where does it end?

3 What kinds of wildlife live on or near the river?

4 Which cities does it flow through?

5 What does the name Mississippi mean?

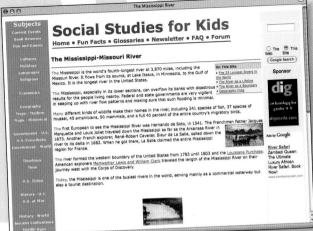

BEFORE YOU READ

1 **LISTENING**

Listen to Chapter Five. Are these sentences true (T) or false (F)?

		T	F
1	Tom and Huck liked Muff Potter.	☐	☐
2	The two boys brought Muff something to drink.	☐	☐
3	There were not many people at the trial.	☐	☐
4	Injun Joe did not go to the trial.	☐	☐
5	Tom told the lawyer the truth about the night in the graveyard.	☐	☐
6	After the trial Muff did not go back to jail.	☐	☐

Muff Potter's Trial

A few weeks later there was Muff Potter's trial. [1] [8] Everyone in St Petersburg talked about it and Tom and Huck were worried.

"Tom," said Huck, "did you tell anyone about... *that*?"

"No, of course not," said Tom. "Poor old Muff, I'm very sorry for him. People say he's a killer, but it's not true. And they'll hang [2] him!"

"I'm sorry for Muff too," said Huck, "He's a kind man. Once he gave me half a fish. But we can't tell anyone about Injun Joe."

"Once he helped me with my kite," said Tom. "Now I want to help him."

1. **trial** : an official meeting when people decide if a person is guilty of a crime or not.

2. **hang** :

46

"Let's go to the jail and take him something to eat," said Huck.

They went to the small jail with some food and saw Muff, who looked sad and tired.

"No one remembers old Muff anymore. But you're my friends and you remember me. Thank you, boys!" said Muff, smiling.

Now Tom felt terrible. He couldn't sleep at night because he thought about poor old Muff in jail.

The day of Muff's trial was a very important one. All the people of St Petersburg went to the trial. Muff was there with his old, dirty clothes and sad face. Injun Joe was there too.

During the trial there were a lot of questions and answers. Things were not going well for poor Muff. Then the lawyer [3] said, "Call Thomas Sawyer!"

Everyone was surprised and looked at Tom. Why was the lawyer calling Tom Sawyer? What did he know? Tom was worried and afraid.

"Thomas Sawyer, where were you on June 17 at midnight?" asked the lawyer.

Tom quickly looked at Injun Joe. "I was in the graveyard," he said.

"Were you near Hoss Williams's grave?" asked the lawyer.

"Yes, sir," answered Tom.

"Why were you there at midnight?" asked the lawyer.

"I went there to see ghosts, with a – a dead cat," said Tom.

A few people started laughing and the lawyer got angry.

"What did you see in the graveyard?" asked the lawyer. "Tell us what happened."

Tom decided to tell the true story. The people at the trial listened to him and were very surprised.

---

3. **lawyer** : a person who advises people on legal problems.

"...and then Muff Potter fell to the ground and Injun Joe took Muff's knife and —"

Suddenly there was a very loud noise. It was Injun Joe, who was jumping out of the window! No one could stop him and he ran away. The sheriff was very angry.

Now everyone knew who killed Dr Robinson and Muff was free. Tom became famous in St Petersburg because he told the truth and saved Muff's life. Tom was happy because he did the right thing, but at night he had terrible dreams about Injun Joe. The hot summer days passed and no one could find Injun Joe.

UNDERSTANDING THE TEXT

1 COMPREHENSION CHECK

Read the paragraphs and choose the best word (A, B or C) for each space. There is an example at the beginning (0).

Tom and Huck talked (**0**) .....about...... Muff Potter's trial because they were worried. They knew he did not kill Dr Robinson but they were afraid (**1**) .... Injun Joe.

The boys liked Muff because he was a kind man. They went to (**2**) .... Muff in jail and brought (**3**) .... some food to eat. Muff was happy (**4**) .... the boys remembered him.

(**5**) .... in the village of St Petersburg went to Muff's trial. Injun Joe was there too. The lawyer asked a (**6**) .... of questions. He called Tom Sawyer and asked him questions about the night of June 17.

Tom decided to tell the truth and the people at the trial (**7**) .... very surprised. Suddenly Injun Joe jumped (**8**) .... of the window and ran away. Tom saved Muff's (**9**) .... and now Muff was free.

0	(A) about	B for	C around
1	A from	B of	C by
2	A look	B watch	C see
3	A he	B him	C his
4	A because	B why	C so
5	A Anyone	B Every	C Everyone
6	A much	B lot	C lots
7	A were	B was	C be
8	A from	B inside	C out
9	A live	B life	C living

2 NOUNS AND ADJECTIVES

A Complete the table with the correct form of the words. Use a dictionary to help you if necessary. There is an example at the beginning (0).

NOUN	ADJECTIVE	NOUN	ADJECTIVE
0 happiness	happy..........	4	angry
1	sad	5 danger	....
2 hunger	....	6	interesting
3	sleepy	7 truth	....

B Now write four sentences about the story using two nouns and two adjectives from the ones above.

T: GRADE 3

3 SPEAKING – JOBS

At the trial there is a lawyer and a sheriff. Talk about the job you want to do. Use these questions to help.

1 What kind of job do you want to do?
2 Why would you like to do this job?
3 Is it an easy, difficult or dangerous job?
4 What subjects do you need to study to do this job?

BEFORE YOU READ

1 VOCABULARY

Match the words from the box with the pictures.

A fireplace   B haunted house   C silver coins   D stone   E hole

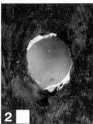

CHAPTER SIX

The House on Cardiff Hill

All young boys want to find a treasure and Tom did, too. One hot summer day he told Huck about his idea.

"That's a great idea!" said Huck happily. "But where can we look for a treasure?"

"Well, robbers put treasures under big trees or in old houses," said Tom. "There's an old house on Cardiff Hill and we can start digging under the big tree near the house. Let's go, Huck!"

It was a hot, sunny day when Tom and Huck walked to Cardiff Hill and started digging under the big tree. They dug for a few hours but found nothing.

"I'm hot and tired," said Tom, "and there's nothing under this tree. Let's go to the haunted house. Nobody lives there and haunted houses sometimes have a treasure."

"The haunted house!" cried Huck. "But haunted houses have ghosts. I don't want to go."

"Ghosts only come out at night," said Tom. "It's daytime now. Come on, Huck,"

"Well, alright," said Huck, who was still afraid.

The haunted house was an old, scary place. They opened the door and went in quietly. They looked around and everything was old and broken. No one lived there. They looked in all the rooms downstairs, but they didn't find any treasure.

Tom and Huck went upstairs and looked around. "Sh!" said Tom.

"What is it?" whispered Huck. "Do you hear ghosts?"

"No," said Tom. "Don't move!"

There were holes in the old wooden floor. Through the holes they could see the rooms downstairs.

"Oh, no!" whispered Tom. "Look! Now there are two men downstairs."

One was an old Spanish man with long white hair and a big hat. The other man was short and wore dirty clothes. They were talking.

"Let's listen to them," whispered Tom.

"It's too hot in here and I'm tired," said the old Spanish man.

When the boys heard his voice they were terrified. "That's not a Spanish man: that's Injun Joe!" whispered Huck. "I know his voice." The boys' faces became white. Injun Joe was dressed as an old Spanish man because he did not want anyone to recognize [1] him.

"What are we going to do with the $650 in silver coins?" asked Injun Joe's friend. "That was a good robbery!"

---

1. **recognize** : when you know who the person or the thing is.

"Well, let's take $30 with us now and hide the bag here," said Injun Joe. "No one knows about this hiding place. We can come back and get it soon."

The short man moved a big stone in the fireplace and pulled out a bag. He took some money from the bag. Then Injun Joe started digging near the fireplace with his knife.

There was a real treasure downstairs and Tom and Huck were excited! Six hundred and fifty dollars was a great treasure for two young boys.

Suddenly Injun Joe stopped digging. "There's something here!" he said excitedly. "I think it's a box." He found an old box and opened it slowly.

"It's money!" cried Injun Joe. "This box is full of gold coins – we're rich!" The two men looked at the gold coins and laughed.

Tom and Huck were happy too.

"This is the treasure of the family that lived here and now it's ours," said Injun Joe, looking at the coins.

"Where can we hide all these coins?" asked the friend. "Can we put the box back under the stone?"

"No, no," said Injun Joe, who was thinking of a better place to hide the box. "This isn't a good place. I know a better place to hide it. We'll wait until it's dark outside and then we'll hide it under the cross. Nobody knows about that hiding place."

When it was dark the two men took the box away. Tom and Huck did not follow them because they were afraid of Injun Joe. But where was the cross? Tom and Huck wanted to find the cross and the big treasure.

UNDERSTANDING THE TEXT

 KET

1 COMPREHENSION CHECK

**Read these sentences about Chapter Six. Choose the correct answer –
A, B or C. There is an example at the beginning (0).**

0 Tom and Huck wanted to dig
- A ☐ under the house on Cardiff Hill.
- B ☑ under the tree near the house on Cardiff Hill.
- C ☐ in the garden of the house on Cardiff Hill.

1 Huck didn't want to go to the house on Cardiff Hill
- A ☐ so he went home.
- B ☐ because he was hot and tired.
- C ☐ because it was a haunted house.

2 Through the holes in the old wooden floor the boys could
- A ☐ see the rooms downstairs.
- B ☐ see ghosts moving around.
- C ☐ hear strange noises.

3 The two men downstairs were
- A ☐ having lunch.
- B ☐ fighting.
- C ☐ looking for something.

4 During their last robbery Injun Joe and his friend stole
- A ☐ a bag full of gold coins.
- B ☐ $650 in silver coins.
- C ☐ a horse.

5 Near the fireplace Injun Joe found
- A ☐ a big knife.
- B ☐ a cross.
- C ☐ an old box.

6 After dark, Injun Joe and his friend
- A ☐ had dinner and slept.
- B ☐ took the treasure to a better hiding place.
- C ☐ hid the treasure behind a stone in the fireplace.

"THIS ISN'T A GOOD PLACE. I KNOW OF A BETTER PLACE TO HIDE IT."

Better is a comparative adjective. We form comparative adjectives like this:

* Adjectives of one syllable, add **-er**, e.g. *tall – taller, old – older*.
* If the adjective is a consonant-vowel-consonant, e.g. *hot*, we double the final consonant, e.g. *hot – hotter, big – bigger*.
* With adjectives of two syllables ending in **-y**, we remove the **y** and add **-ier**, e.g. *easy – easier, happy – happier*.
* With adjectives of two or more syllables, we use **more** before the adjective, e.g. *beautiful – more beautiful, important – more important*.
* Be careful! Some adjectives have an irregular comparative forms, e.g. *good – better, bad – worse*.

2 COMPARATIVE ADJECTIVES

Fill in the gaps with the correct comparative adjective.

1 Mary is pretty but Becky is .... (*pretty*)

2 The haunted house is .... (*big*) than Tom's house.

3 Huck is short but Injun Joe's friend is .... (*short*)

4 Life on Jackson's Island was .... (*interesting*) than going to school.

5 It was .... (*easy*) to sit under a tree than to paint the fence.

6 The apple was good but the chocolate cake was .... (*good*).

7 During the fight Tom was .... (*strong*) than the big boy.

3 WORD GAME

Unscramble each of these words. Write the letters in the boxes. Then copy the numbered letters to the boxes with the same numbers to discover the mysterious words.

1 ehsuo   [2] [ ] [U] [ ] [ ]

2 sretuare   [1] [R] [ ] [ ] [U] [ ] [3]

3 ltsoceh   [4] [ ] [L] [6] [ ] [ ] [ ]

4 rayewl   [ ] [ ] [ ] [Y] [ ] [5]

5 sthgo   [ ] [ ] [ ] [7] [T]

6 hocosl   [8] [C] [ ] [ ] [ ] [L]

Mystery words [ ] [ ] [ ]   [ ] [ ] [ ] [ ] [ ]
                    1 2 3   4 5 6 7 8

57

A view of the River Missouri.

Missouri

  Tom Sawyer and his friends lived in the American state of Missouri. The word Missouri comes from the name of an American Indian people called Missouria which meant "town of the large canoes." [1] Seven American Indian tribes lived in Missouri before the European explorers came. They were the Chickasaw, Illini, Ioway, Missouria, Osage, Oto and Quapaw tribes.

During the late 1600s French explorers traveled down the Mississippi River and explored this big territory. [2] They called it Louisiana after King Louis XIV of France. In 1764 the French

1. **canoe** :
2. **territory** : a very big piece of land.

explorer Pierre Laclède founded [3] the city of St Louis where the Mississippi and Missouri Rivers meet. The city grew quickly because of its position on the two great rivers.

In 1803 the American President Thomas Jefferson bought Louisiana from France for $15 million. He asked Meriwether Lewis and William Clark to explore this big territory. The famous Lewis and Clark Expedition started near St Louis and traveled about 8,000 miles (12,875 kilometers) from Missouri to the Pacific Ocean. Lewis and Clark made maps and brought back a lot of important information about the territory.

After the Lewis and Clark Expedition people knew more about the new territory and thousands of pioneers [4] went to live in the West. They built homes, towns and cities and America grew quickly. These pioneers began their long journey in the town of Independence, Missouri. Soon Independence was called the "Gateway to the West".

In 1821 Missouri became the 24th state of the United States and today almost six million people live there.

Indians guiding Lewis and Clark on their expedition.

3. **founded** : created, built for the first time.
4. **pioneers** : people who go to live in a new land for the first time.

1 COMPREHENSION CHECK
Are the following sentences true (T) or false (F)? Correct the false ones.

T F

1 Missouria was the name of an American Indian people. ☐ ☐
2 Only the Chickasaw people lived in Missouri. ☐ ☐
3 King Louis XIV of France explored the Mississippi River in the 1700s. ☐ ☐
4 St Louis is a city on the Missouri and Mississippi Rivers. ☐ ☐
5 Meriwether Lewis and William Clark were two famous American explorers. ☐ ☐
6 The pioneers began their long journey to the West in St Louis. ☐ ☐

PROJECT ON THE WEB

Let's find out more about Missouri!

Connect to the Internet and go to www.blackcat-cideb.com. Insert the title or part of the title of the book into our search engine. Click on the internet project link. Click on the relevant link for this project.

Missouri is an important American state because of its history and geographic position. It is called the "Show Me State" because there are 50 many things to see.

Work with a partner and answer these questions about Missouri.

1 What is the capital of Missouri and what is the largest city?

2 What are the major rivers?

3 What American President was born in Missouri?

4 What is the state flower and the state bird?

5 What are the most important industries?

6 Which American states border Missouri?

7 What is the St Louis Gateway Arch?

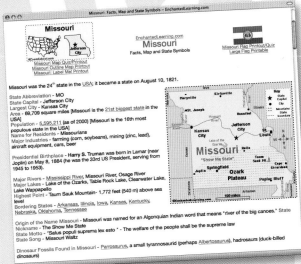

BEFORE YOU READ

1 **VOCABULARY**

Match the words from the box with the pictures.

candle   bat   cave

1

2

3

KET

2 **LISTENING**

 Listen to the first part of Chapter Seven and choose the correct answer – A, B or C.

1   What did the boys and girls eat at lunchtime?

A

B

C

2   What did everyone take inside McDougal's Cave?

A

B

C

3   How did the children go back to St Petersburg?

A

B

C

Lost in the Cave

On Saturday it was Becky Thatcher's birthday and all her friends were excited.

"Tom, you're invited to my birthday picnic near the river," said Becky happily. "Please come! All my friends will be there and we'll have lots of fun. After the picnic we can visit McDougal's Cave."

"Thanks, Becky!" said Tom. "It sounds great." He was very happy because he liked Becky. He stopped worrying about Injun Joe and the treasure and thought about Becky's birthday picnic.

On Saturday morning a big boat took Becky, Tom and their friends down the river. There were no mothers or fathers, but only a few older boys and girls who were about eighteen years old. The boys and girls played games and at lunchtime everyone ate good food and the big birthday cake.

In the afternoon the children went to visit McDougal's Cave. It was an exciting place but it was a bit scary. Everyone had candles because it was very dark inside.

McDougal's Cave was very big and had hundreds of long tunnels. [1] It was easy to get lost. The children played and ran around in the big cave, but they always stayed near the entrance. They did not want to get lost. Tom and Becky wanted to find new tunnels inside the cave. They walked and walked until they were alone. Where were the other children? They were lost!

In the evening the other children got on the boat and went back to St Petersburg. They talked and laughed, but they were very tired. They did not see that Tom and Becky were not on the boat. /END

Huck did not know about the picnic. The mothers of the village did not like him and never invited him to birthday picnics.

That evening Huck had a clever plan. He wanted to find Injun Joe's treasure. He hid behind a tree outside an old house. He thought, "I'll stay here and wait. When Injun Joe comes out I'll follow him and I'll find the treasure. I'll tell Tom tomorrow."

It was a cold night and it started raining. At midnight two men came out of the old house: Injun Joe and a friend. Huck followed them quietly.

"How strange! They're going to Widow [2] Douglas's house," thought Huck. "But why?"

Suddenly the two men stopped. Injun Joe said, "Many years ago Widow Douglas's husband did something bad to me. Now I want to hurt his widow. I'm going to cut her ears, her nose and her face! And you have to help me!"

1. **tunnels** : long, dark roads or paths under the ground.
2. **widow** : her husband is dead.

"Oh, please don't hurt the poor old woman," said his friend. But Injun Joe laughed and walked away.

Huck heard the conversation and wanted to run away. But he remembered that Widow Douglas was always kind to him. "I must help her," thought Huck. "Those men are dangerous."

Huck had an idea and ran to Bill Welsh's house. "Mr Welsh, it's me, Huck! Open the door!"

Mr Welsh opened the door.

"Please help me, Mr Welsh," said Huck. "Two men want to hurt Widow Douglas!"

Mr Welsh and his sons ran to the widow's house and surprised Injun Joe and his friend. When the two men saw Mr Welsh and his sons they ran away.

"Thank you for helping me," said Widow Douglas to everyone.

The next day Huck went to see Mr Welsh.

"You're brave, [3] Huck," said Mr Welsh. "You saved Widow Douglas's life. Injun Joe and his friend are terrible men. We have to find them. Now sit down and have some breakfast with my family."

Huck was happy because he saved Widow Douglas's life. And now he had new friends, the Welsh family.

No one in St Petersburg knew where Tom and Becky were. Tom and Becky's families were very worried.

Tom and Becky were in McDougal's Cave. Tom tried to find the entrance to the cave but he couldn't. They spent the night there and the next morning they did not know what to do. They were cold, hungry and afraid. He took Becky's hand and they walked in the long tunnels for hours with their candles.

3. **brave** : not afraid.

Then they found a big empty space with a lot of black bats.

"Oh no!" cried Becky. "Bats! They're flying everywhere! I'm scared, Tom!"

The bats flew over their heads and when Tom and Becky ran away the bats followed them. At last the bats went away.

"Tom, where are we and what are we going to do?" asked Becky, who was crying.

"I'm sorry, Becky, I don't know," said Tom sadly.

They walked in the long, dark tunnel: they were tired and hungry.

"There are too many tunnels here," said Becky, crying. "No one will ever find us. We're going to die here!"

"Don't worry, Becky," said Tom bravely, "we'll get out of this cave." Becky had a piece of cake in her pocket and they ate it. Then their candles died and everything was dark. They were tired and slept.

When they woke up they were very hungry and thirsty. Tom heard a noise.

"Becky, did you hear a noise?" asked Tom. "Someone is looking for us!"

Becky looked at Tom and smiled. "I'm going to see," said Tom. "Stay here, Becky! Don't move!"

UNDERSTANDING THE TEXT

1 **COMPREHENSION CHECK**
Are these sentences true (T) or false (F)? Correct the false ones.

T   F

1   Becky Thatcher was going to have a birthday picnic in McDougal's Cave. ☐ ☐
2   A big boat took the all the young people down the river. ☐ ☐
3   The children who visited McDougal's Cave had candles. ☐ ☐
4   Most of the children who played in the cave got lost. ☐ ☐
5   Tom and Becky were not on the boat because they were still in the cave. ☐ ☐
6   Huck was not invited to Becky's birthday. ☐ ☐
7   Injun Joe wanted to hurt old Widow Douglas. ☐ ☐
8   Huck asked Bill Welsh and his sons to find Tom and Becky. ☐ ☐
9   Becky and Tom were not afraid of the bats. ☐ ☐
10 Tom heard a noise and went to see who it was. ☐ ☐

 KET

2 **WRITING**
Complete Becky's letter. Write ONE word for each space. There is an example at the beginning (0).

Dear Susan,

It's (0) ......my...... birthday next Saturday and I'm having
(1) .... big birthday party on the river. My mother and
father planned (2) .... and I'm very happy.
Please come (3) .... my party. All of my friends will
(4) .... there. You can meet Tom Sawyer, who is a boy
(5) .... like. He's very friendly and he draws well too.
There (6) .... many games to play (7) .... the morning.
Then there's a big lunch and the birthday cake. Later in the day
we can visit McDougal's Cave. Do (8) .... like caves? They're
a bit scary but they're fun.
Meet me (9) .... half past ten at the river boat on Saturday.
Tell (10) .... brother Jim to come too!
    Your friend, Becky

67

KET

3 LISTENING – BATS

Listen to the facts about bats. For questions 1-5 tick (✓) A, B or C.

1 What are bats?
   A ☐ They are birds that fly at night.
   B ☐ They are animals that fly at night.
   C ☐ They are animals that sleep all night.

2 How many kinds of bats are there in the United States?
   A ☐ 40
   B ☐ 400
   C ☐ 1,100

3 Where do bats usually live?
   A ☐ in rainy places
   B ☐ in warm, dark places
   C ☐ in tall trees

4 Which kinds of bats are dangerous?
   A ☐ fruit bats and vampire bats
   B ☐ flying fox bats
   C ☐ American brown bats

5 How long are the flying fox's wings?
   A ☐ about 31 centimeters
   B ☐ about 100 centimeters
   C ☐ about 150 centimeters

6 How do bats find their way?
   A ☐ They use their eyes and ears.
   B ☐ They use their wings and ears.
   C ☐ They use their eyes and wings.

7 Why is a bat cave noisy?
   A ☐ because it's always windy inside caves
   B ☐ because bats like to fight
   C ☐ because bats make
         sounds to talk to each other

BEFORE YOU READ

KET

1 LISTENING

13 Listen to part of Chapter Eight and choose the correct answer – A, B or C.

1 Why couldn't Injun Joe see Tom?
   A ☐ Tom was hiding.
   B ☐ Injun Joe was sleeping.
   C ☐ It was dark.

2 What happened to Becky's mother?
   A ☐ Her hair became white.
   B ☐ She became ill.
   C ☐ She got lost in the cave.

3 When did Tom and Becky go back to St Petersburg?
   A ☐ on Tuesday morning
   B ☐ on Thursday night
   C ☐ on Tuesday night

4 What did Tom use to help him in the cave?
   A ☐ a candle
   B ☐ a piece of string
   C ☐ a key

5 Who has the keys to the doors in front of the cave entrance?
   A ☐ Injun Joe
   B ☐ Huck
   C ☐ Becky's father

6 What happened to Injun Joe?
   A ☐ He died inside the cave.
   B ☐ He ran away with the treasure.
   C ☐ He went to jail.

"Really?" asked Huck, who was excited.

"Yes," said Tom, "Injun Joe hid the money there."

"That's wonderful, Tom!" cried Huck.

"The money's in the cave and we can get it," said Tom happily.

"But how?" asked Huck. "Now the entrance is closed with big doors and it's easy to get lost in the cave."

"I found another entrance," said Tom. "And we won't get lost because I've got candles, two spades [3] and a long string."

They took a small boat and went down the Mississippi River.

"Look, Huck," said Tom, "here's the other entrance."

"It's very small," said Huck. "That's why nobody knows about it."

Tom and Huck went into the cave. They had candles and used the long string to help them.

Tom suddenly stopped and said, "This is where I saw Injun Joe."

"Oh," said Huck, "his ghost is probably here."

"Don't worry about his ghost," said Tom. "We have to find the hiding place."

"Alright," said Huck, who was afraid of ghosts, "but let's hurry."

Tom looked around slowly and said, "I saw Injun Joe standing here." Then he cried, "Look, here's the cross!" They looked at a large black cross.

"This is the cross that Injun Joe was talking about," said Huck, "but where's the treasure?"

"Injun Joe said *under* the cross," said Tom. "Let's dig under the cross."

They dug and found some big wooden boards. [4] They moved the boards to one side and found a small room.

3. spade :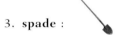
4. boards : (here) flat, rectangular pieces of wood.

There was a small bed, some old candles and a few bottles.

"Look, near the bed," cried Tom. "It's the treasure box!" The boys opened it and inside they found the gold and silver coins.

"We're rich!" cried Huck happily. "We found the treasure!"

"This is wonderful!" cried Tom, looking at the coins. "Let's take the treasure and get out of this cave!"

They took the heavy treasure box and followed the long string to the small entrance. They got into the boat and traveled up the river. When they got to St Petersburg they took the treasure to Aunt Polly's house. When the people in St Petersburg saw the boys and the treasure box, they followed them.

"Oh, Tom! Huck!" cried Aunt Polly. "What are you doing? What's in that old box?"

"Wait and see, Aunt Polly," said Tom happily.

Tom opened the treasure box and everyone was amazed. [5] There was $12,000 dollars!

"Half of the treasure is Huck's and half is mine!" said Tom, smiling at his friend.

Then Tom told his long story about Injun Joe and the treasure and everyone listened to him. Now Tom and Huck were the richest people in St Petersburg!

5. **amazed** : very, very surprised.

UNDERSTANDING THE TEXT

1 COMPREHENSION CHECK

Are these sentences "Right" (A) or "Wrong" (B)? If there is not enough information to answer "Right" or "Wrong", choose "Doesn't say" (C). There is an example at the beginning (0).

0 Injun Joe was carrying a candle.
   (A) Right   B Wrong   C Doesn't say

1 Tom was scared because Injun Joe called him.
   A Right   B Wrong   C Doesn't say

2 Becky was not feeling well.
   A Right   B Wrong   C Doesn't say

3 More than fifty people from St Petersburg went to McDougal's Cave to look for Tom and Becky.
   A Right   B Wrong   C Doesn't say

4 Tom and Becky returned to St Petersburg at ten o'clock on Tuesday night.
   A Right   B Wrong   C Doesn't say

5 Tom didn't get lost in the cave because he used a long string to help him.
   A Right   B Wrong   C Doesn't say

6 Mr Thatcher was the only person who had the keys to the doors of the cave.
   A Right   B Wrong   C Doesn't say

7 Injun Joe died the day before they found him.
   A Right   B Wrong   C Doesn't say

8 Tom and Huck started digging under the black cross and got lost in the cave.
   A Right   B Wrong   C Doesn't say

2 VOCABULARY

Read the descriptions. What is the word for each one? The first letter is already there. There is one space for each other letter in the word.

0 not strong   w _e_ _a_ _k_

1 very surprised   a _ _ _ _ _

2 something you can use to make light   c _ _ _ _ _

3 something you can use to tie things   s _ _ _ _ _

4 something you use to open a locked door   k _ _

5 something you use to dig   s _ _ _ _

2 PICTURE SUMMARY

Look at the pictures from *The Adventures of Tom Sawyer* below. They are not in the right order. Put them in the order in which they appear in the story.

 A

 B

 C

 D

 E

 F

G

....

....

H

....

....

I

....

....

J

....

....

K

....

....

L

....

....

3 **A GRAPHIC NOVEL**

Photocopy these two pages, cut out the pictures and stick them on paper in the right order. Think of words to put in speech or thought bubbles to show what the characters are saying or thinking. Do not use the words that were used in this book! Then write at least one sentence under each picture to narrate what is happening.

This reader uses the **EXPANSIVE READING** approach, where the text becomes a springboard to improve language skills and to explore historical background, cultural connections and other topics suggested by the text.

The new structures introduced in this step of our **G**REEN **A**PPLE series are listed below. Naturally, structures from lower steps are included too.

The vocabulary used at each step is carefully checked against vocabulary lists used for internationally recognised examinations.

 # Step 1 A1

Verb tenses
Present Simple
Present Continuous
Future reference: Present Continuous; *going to*;
  Present Simple

Verb forms and patterns
Affirmative, negative, interrogative
Short answers
Imperative: 2nd person; *let's*
Infinitives after some very common verbs (e.g. *want*)
Gerunds (verb + *-ing*) after some very common verbs
  (e.g. *like*, *hate*)

Modal verbs
Can: ability; requests; permission
Would ... like: offers, requests
Shall: suggestions; offers
Must: personal obligation
Have (got) to: external obligation
Need: necessity

Types of clause
Co-ordination: *but*; *and*; *or*; *and then*
Subordination (in the Present Simple or Present
  Continuous) after verbs such as: *to be sure*; *to know*;
  to think; *to believe*; *to hope*, *to say*; *to tell*
Subordination after: *because*, *when*

Other
Zero, definite and indefinite articles
Possessive *'s* and *s'*
Countable and uncountable nouns
Some, any; *much, many, a lot*; *(a) little, (a) few*;
  all, every; etc.
Order of adjectives

Available: